Wonder Wings

GUESS WHO'S FLYING

WRITTEN BY
REBECCA E. HIRSCH

ILLUSTRATED BY
SALLY SOWEOL HAN

ABRAMS BOOKS FOR YOUNG READERS
NEW YORK

The art in this book was created with gouache, watercolors, color pencils, and soft pastels.

Cataloging-in-Publication Data has been applied for and may be obtained from the Library of Congress.

ISBN 978-1-4197-6925-2
eISBN 979-8-88707-088-9

Book design by Heather Kelly

Printed and bound in China
10 9 8 7 6 5 4 3 2 1

ABRAMS The Art of Books
195 Broadway, New York, NY 10007
abramsbooks.com

To Mom, with love
—R.E.H.

To every child reading this book, may you spread your wonderful wings as you grow
—S.S.H.

Wings can soar and wings can skim.

Wings can hover, hum, and swim.

Small wings, big wings, near and far.

Can you guess whose wings these are?

Wings can buzz and pollinate,

weighted with delicious freight.

Then back to hive, to hollow tree.

These wings are on a . . .

HONEYBEE!

Wings can cling to leafy trees,
waiting for a hearty breeze.
Twirling
down,
they land in weeds.
These wings are on a . . .

MAPLE SEED!

Wings can make a thrumming sound
in gardens as they race around.
Beating fast, these wings look blurred.
These wings are on a . . .

HUMMINGBIRD!

Wings can sing a treetop tune
on summer nights beneath the moon.
They trill *she didn't.* Then *she did.*
These wings are on a . . .

KATYDID!

Wings skim over fields in bloom
and chase a trail of sweet perfume.
Drifting low—lifting high.
These wings are on a . . .

BUTTERFLY!

Wings can circle ponds and pools.

They glint and gleam like sunlit jewels,

fast and fleet as they flash by.

These wings are on a . . .

DRAGONFLY!

Wings can swoop in twilight air.
The hunt is on, so bugs beware!
Watch out beetles, moths, and gnats.
These wings are on a . . .

BIG BROWN BAT!

Wings can wear an oily sheen,
a dab of grease that keeps them clean,
sealed from water, mud, and muck.
These wings are on a . . .

MALLARD DUCK!

Atop a hill of ice and snow,
wings can't fly, but they can go—
tobogganing, a wintry trick!
These wings are on a . . .

PENGUIN CHICK!

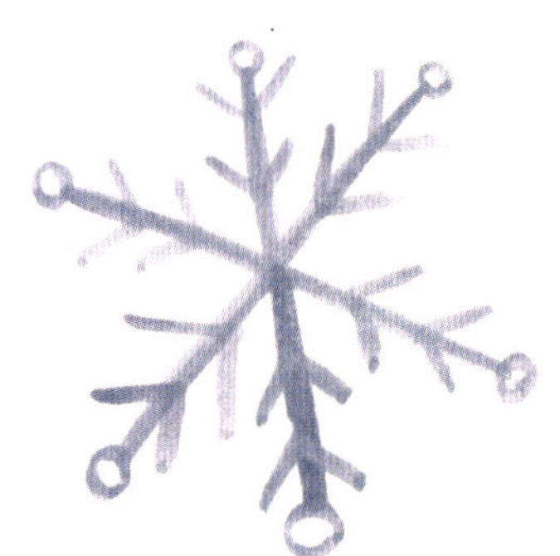

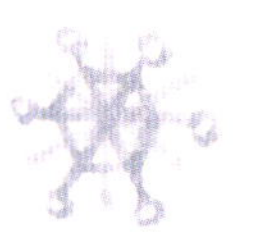

Above the dark and quiet ground,

wings drift by without a sound.

A hunter on a midnight prowl.

These wings are on a . . .

GREAT HORNED OWL!

Wings can even be colossal!

No more skin and bone, just fossil.

Who used these giant wings before?

These wings were on a . . .

PTEROSAUR!

Wings can also be for you!

Buckle up and meet the crew.

See clouds go past the windowpane.

These wings are on a . . .

JET AIRPLANE!

HOW THESE WINGS WORK

HONEYBEE

Honeybee wings are small but mighty. A bee has four wings, two on each side of its body. Each pair of wings zips together, creating a large surface that aids in flight. Bees beat their wings front to back. Air rushing past the wings generates lift, an upward force. This raises the bee into the air. By using short flaps instead of long ones, bees create enough lift to carry heavy loads of pollen and nectar back to the hive.

MAPLE SEED

Maple trees begin life as seeds with papery wings. The winged seeds dangle in pairs from tree branches. You might know them as helicopters or whirlybirds. When the wind blows, the pairs split apart and spin in the air. As each seed spins, a spiral of air forms on top of the wing. This tiny tornado sucks the seed upward, giving it lift. Staying airborne allows the seed to travel to a new place to grow far from the mother tree.

HUMMINGBIRD

Ruby-throated hummingbirds weigh about as much as one or two pennies. Being light helps them overcome Earth's gravity. These tiny birds beat their wings fifty to seventy times a second! The wings are curved on top and flat underneath. Air flows faster over the curved top of the wing and slower beneath the flat bottom. Air pressure builds up under the wing. This lifts the bird into the air. Hummingbirds flap their wings forward and backward in a figure eight. This movement creates lift on both the front and back stroke.

KATYDID

Katydids communicate with their wings. They sing by rubbing a row of tiny teeth on one wing against a scraper on the other. This vibrates the air and creates a raspy sound. Each species of katydid makes its own song. On summer and fall nights, you may hear one species, the common true katydid, singing in the trees. Their song sounds like "Katydid . . . she didn't . . . she did." Katydids live on trees, bushes, or grasses, and their wings are camouflaged to blend in with leaves. So, you are more likely to hear a katydid than see one.

BUTTERFLY

Butterflies soar on big, colorful wings. They flap their wings slowly, about ten times per second. The wings bend and flex with each flap. Butterflies beat their wings in a figure-eight pattern, similar to the way hummingbirds do. At the top of each wingbeat, the wings clap together, forming a pocket of air. This creates a jet of air that helps propel the insect forward. Thanks to their big wings, butterflies can take off and turn quickly, making them tricky for hungry predators to catch.

DRAGONFLY

Dragonflies are acrobats of the air. They can zip forward and screech to a stop in midair. They can sweep from side to side, do a barrel roll, or even fly backward. Dragonflies zoom about on four transparent wings. Their wings can move in different directions at the same time. This allows them to perform their daredevil moves, which they use to hunt flying insects and to engage in aerial battles with other dragonflies.

BIG BROWN BAT

Bats are the only mammals that can fly. Their wings are made of a flexible membrane. This thin layer of tissue stretches from the fingers of a bat's hand to the ankles on its back legs. Bats maneuver in the air by bending and flexing their fingers and legs, which changes the shape of their wings. Some of their wing bones are bendable, which helps the wings flex even more. This gives bats the ability to quickly switch directions as they chase down insects in the air.

MALLARD DUCK

Mallards are waterbirds. You may see them tipping forward in lakes and streams as they look for food. Even when these ducks swim, they stay dry. Their wings and bodies are waterproof. This is because mallards spend a lot of time preening. With their beaks, they dab oil from a gland near their tail and smear it on their feathers. Oil and water don't mix, so preening prevents these water birds from getting wet. The structure of the feathers also helps seal out water, and oil keeps the feathers in good shape.

GREAT HORNED OWL

Great horned owls swoop silently to sneak up on prey. Like other owls, their wings are large compared to their bodies. That means owls can glide without having to flap much. Flapping stirs the air and produces sound. Gliding is quieter. Owl wings have a comblike edge on top. The bottom edge has a soft fringe. These edges break up the air rushing past into smaller streams. This muffles the sound and makes their wings as quiet as a whisper.

PENGUIN CHICK

Penguin wings are called flippers. These birds don't fly, but they do use their flippers to get around. From a young age, penguins flop onto their bellies and slide forward on the ice. They steer and push with their flippers and feet. This move, called tobogganing, is a much faster way to cross the ice than walking. Grown penguins also swim with their wings. They propel themselves through the water by flapping their flippers.

PTEROSAUR

Pterosaurs were ancient flying reptiles. They ruled the skies in the age of dinosaurs. There were over two hundred kinds of pterosaurs. The smallest had a wingspan smaller than a big brown bat's. The largest had a wingspan bigger than an F-16 fighter jet. Their wings were made of a stretchy membrane that reached from their back legs to their fourth finger (they had no fifth finger). Like bats, pterosaurs moved their fingers and legs to adjust the shape of their wings.

JET AIRPLANE

We humans don't have wings of our own, but we have invented winged things to help us fly. Like a bird's wings, airplane wings are curved on top and flat underneath. When a plane speeds down the runway, the air rushes faster over the top curve. The air flows under the flat bottom more slowly. This gives the air beneath the wing higher pressure than the air above the wing and—liftoff!

SELECT SOURCES

Agosta, Salvatore J. "Habitat use, diet and roost selection by the Big Brown Bat (*Eptesicus fuscus*) in North America: a case for conserving an abundant species," *Mammal Review* 32, no. 3 (September 2002): 179-198.

Fei, Yueh-Han John, and Yang, Jing-Tang. "Importance of body rotation during the flight of a butterfly." *Physical Review E* 93, no. 3 (March 2016). journals.aps.org/pre/abstract/10.1103/PhysRevE.93.033124.

Horton, Jennifer. "How Do Ducks Float?" HowStuffWorks.com. animals.howstuffworks.com/birds/duck-float1.htm.

Jernigan, Christopher M. "How Do Bees Fly?" Ask A Biologist. Arizona State University. askabiologist.asu.edu/how-do-bees-fly.

Lentink, D., W. B. Dickson, J. L. van Leeuwen, and M. H. Dickinson. "Leading-Edge Vortices Elevate Lift of Autorotating Plant Seeds." *Science* 324, no. 5933 (June 2009): 1438. www.science.org/doi/10.1126/science.1174196.

Ogden, Lesley Evans. "The Silent Flight of Owls, Explained." Audubon. July 28, 2017. www.audubon.org/news/the-silent-flight-owls-explained.

"Pterosaurs: Flight in the Age of Dinosaurs." American Museum of Natural History. January 4, 2015. www.amnh.org/exhibitions/pterosaurs-flight-in-the-age-of-dinosaurs.

Rüppell, Georg. "Lords of the Air." Chapter 5 *Dragonflies of the World* by Jill Silsby. Collingwood, Victoria, Australia: CSIRO, 2001.

Shaw, Robert J., ed., "Dynamics of Flight." NASA. May 13, 2021. www.grc.nasa.gov/www/k-12/UEET/StudentSite/dynamicsofflight.html.

"Species *Pterophylla camellifolia* - Common True Katydid," BugGuide.net. bugguide.net/node/view/7075.

Than, Ker. "Why Bats Are More Efficient Flyers Than Birds." Live Science. January 22, 2007. www.livescience.com/1245-bats-efficient-flyers-birds.html.

Weidensaul, Scott, Tara Rodden Robinson, Robert R. Sargent, Martha B. Sargent, and Theodore J. Zenzal. "Ruby-throated Hummingbird (*Archilochus colubris*)." Birds of the World. Cornell Lab of Ornithology. March 4, 2020. birdsoftheworld.org/bow/species/rthhum/cur/introduction.

"Why Do Emperor Penguins Toboggan?" Woods Hole Oceanographic Institution. www.whoi.edu/know-your-ocean/did-you-know/did-you-know-why-do-emperor-penguins-toboggan.